Management Division 2021

Held at the 2021 AIChE Annual Meeting

Boston, Massachusetts, USA and Online
7 - 11 November and 15 - 19 November 2021

ISBN: 978-1-7138-5701-3

Printed from e-media with permission by:

Curran Associates, Inc.
57 Morehouse Lane
Red Hook, NY 12571

AIChE
120 Wall Street, FL 23
New York, NY 10005-4020

Phone: (800) 242-4363
Fax: (203) 775-5177

www.aiche.org

Additional copies of this publication are available from:

Curran Associates, Inc.
57 Morehouse Lane
Red Hook, NY 12571 USA
Phone: 845-758-0400
Fax: 845-758-2633
Email: curran@proceedings.com
Web: www.proceedings.com

TABLE OF CONTENTS

637362 The Use of Digital Technologies in Materials and Process Design

Abstract:

The topic of Digitization in the Chemical and Materials Industry will be discussed. Today scientists and engineers are facing more complex multidisciplinary problems. The proliferation of the Internet of Things that is, inexpensive instrumentation connected via high-bandwidth and low-latency, have given engineers working in manufacturing access to tremendous amount of data. A similar situation is happening to engineers working in research and development conducting data-intensive experiments and simulations. Data Science is one of the digital technologies in the 3M Periodic Table of Technology Platforms. This talk will include examples on how Data Science in combination with materials and processing technologies is being used successfully.

637363 The Digital Transformation of the Chemicals Industry - and How to Prepare

Abstract:

Over the course of the next 20 years, the chemicals industry will see more change than it has in the last 50, in large part due to digital technologies. Digital transformation is affecting business processes from R&D, to manufacturing, to supply chain, to sales and go-to-market -- and the data from all these digital efforts is tying those diverse functions to optimize the chemical enterprise and create new business models and sources of value. This presentation looks at how the industry will change through 2040 and examines what the new sources of competitive advantage will be for both companies and individuals.

635042 Award Presentation to Senyo Opong

Abstract:

Formal presentation of 2021 Management Award to Dr. Senyo Opong of DuPont

635041 Developing and Sustaining Critical Engineering Knowledge and Expertise in a Rapidly Changing World

Abstract:

Over the past few years, we have witnessed the attrition of deep technical expertise and subsequent loss of technical knowledge in many companies as a result of portfolio shifts, mergers and acquisitions, headcount reductions and retirements. This talk will focus on our strategy to sustain critical knowledge and competencies in the midst of a rapidly changing corporate environment, and our approach to building the deep technical expertise needed to drive innovation and support the safety and profitability of our manufacturing assets. I will discuss how we develop a talent pipeline that enables diversity equity and inclusion, how we coach and mentor early career employees, and how we develop and sustain technical capability and manage critical knowledge through competency teams and communities of practice.

636185 Lessons from the Renewal of a Corporate Technology Organization

Abstract:

Technology organizations in the chemical industry and affiliate industries will continue to evolve and require renewal over the next decade, challenging technology leaders to adapt and develop effective hiring, retention and career development strategies for their engineers and scientists. This talk will share learnings and recommendations from my experiences gained through the creation of The Chemours Company as a spin-off chemical company, with a focus on maintaining core competencies and renewing a global engineering organization.

636131 Chemical Engineers in the Pharmaceutical Industry

Abstract:

Chemical engineers in the pharmaceutical industry work across a multitude of roles spanning discovery, research and development, manufacturing and operations, data analytics, and supply chain. At the same time, the diversity of molecular therapeutics has expanded beyond the traditional small molecules and proteins to include oligonucleotides, antibody-drug conjugates, cell and gene therapies and vaccines (made through many new platforms: viral vectors, mRNA, protein subunits). The delivery of these therapies occur through many different means (oral, intravenous, subcutaneous, inhalation, nasal, ocular, intramuscular, etc.). Given the range of industries chemical engineers can work in (oil/energy, commodity/specialty chemicals, materials, consumer products), it is highly competitive to attract chemical engineers to fulfill the industry's need for expertise, innovation, and talent. This talk will discuss some approaches to attract chemical engineers to the industry, to develop their expertise, and to grow them into leaders for the future.

635318 Management an easy mountain to climb? … really?

Abstract:

Mountain Climbing and Management have a lot in common. Both 'fields' present many difficulties but are not insurmountable although the required preparations must not be underestimated. With the right tools and attitude success can be achieved in both endeavors. Conditions that must be assessed have always changed in the past, are changing at this very moment and will certainly continue to change in the future!! No reason to be deterred. In the end it is always a question of how to best deal with people, their capabilities, differences and emotions. In this presentation we will jointly look at some of these underlying principles.

636188 Q & A of all speakers

Abstract:

The chairs will preside over a 10 minute general Q&A and discussion involving all session speakers and audience.

631529 Multiscale Modeling of Spray Coating of Perovskite QDs: Understanding the Role of Molecular Interactions in Particle Aggregation.

Abstract:

The US Department of Energy has set high-priority targets to reduce the cost of photovoltaics from ~ 10 cents/kWh to half of that at 5 cents/kWh by 2030. This has fueled a plethora of research in the discovery, optimization, and commercial production of high-efficiency solar cells. Perovskite solar cells (PSCs), especially $CsPbBr_3$, have received special attention due to their high quantum efficiency and tunable absorption/emission spectra [1]. Often, the perovskite quantum dots (QDs) are synthesized using one of the various batch synthesis routes, and after appropriate post-processing, they are spin-coated and separately cured to make a functional layer for the PSCs [2]. However, the problems of batch-to-batch variation, low-material utilization, and separate curing required in the spin coating process have slowed down and made the commercial production of PSCs difficult. In response, recent studies have demonstrated that spray coating can counter these challenges associated with spin coating by (a) providing a fast-scalable continuous production alternative with better material utilization; (b) offering a large control space for uniform deposition without the need for a separate curing step [3], and (c) integrating it with a continuous crystallizer to result in higher production capacity [4]. That being said, modeling and control of spray coating is a non-trivial task due to a combination of macroscopic phenomena (e.g., transport and thermal), and molecular-level interactions (e.g., dipole-dipole, van der Waals, and steric forces), thereby requiring a multiscale modeling approach.

In spray coating, the colloidal solution from the continuous crystallizer is atomized and sprayed onto a thermally heated silicon substrate through a pressurized nozzle. Consequently, the atomized droplets impinge onto the substrate and form a liquid splat followed by evaporation of the solvent (toluene). In this work, the spray coater is modeled using a system of macroscopic mass and energy balance equations (MEBE), which describe the effect of surface tension, vapor pressure, nozzle velocity, and the size of the impinging droplet on solvent evaporation. Further, during the solvent evaporation, the QD particles undergo particle aggregation due to the presence of various molecular interactions, and can lead to the formation of a non-uniform film, and demonstrate the coffee-ring effect. Thus, a modified form of the discrete element method (DEM), which solves newton's equation of motion for each QD particle, is utilized to accurately model these microscopic interactions [5]. Specifically, the DEM considers four molecular interactions, viz., steric hindrance, dipole-dipole forces, Brownian motion, and van der Waals forces. This is because, (a) the QD particles have a long-chain ligand-shell around them, which results in steric hindrance (repulsion) if two QD particles come very close to each other during aggregation; (b) experimental observations show a non-zero dipole moment for the QD particles [6]; and (c) the QD particles have a size range of 5-10 nm, which makes them very susceptible to undergo Brownian motion. Finally, the continuum model (MEBE) and the DEM are dynamically coupled to develop a high-fidelity multiscale model of spray coating.

The simulation results generated for different nozzle velocities and QD concentrations demonstrate the typical coffee ring formation as seen in the experiments [7]. The results also elucidate the effect of controllable macroscopic variables (i.e., substrate temperature, nozzle velocity, and QD particle concentration) on microscopic film evolution (i.e., aggregate formation, fractal structure and size). Overall, the proposed multiscale modeling framework provides a mechanistic understanding of the role of molecular interactions in QD particle aggregation and film deposition during temperature-controlled spray coating of QD particles for application in PSCs. Furthermore, this work provides a general platform for multiscale modeling of spray coating of different particle systems and will be incorporated with advanced optimization or process control frameworks to ensure adequate set-point (film thickness and smoothness) tracking in future works.

Literature Cited:

1. Protesescu L, Yakunin S, Bodnarchuk MI, Krieg F, Caputo R, Hendon CH, Yang RX, Walsh A, Kovalenko MV. Nanocrystals of cesium lead halide perovskites ($CsPbX_3$, X= Cl, Br, and I): novel optoelectronic materials showing bright emission with a wide color gamut. Nano Letters. 2015 Jun 10; 15(6): 3692-6.
2. Chen LC, Tien CH, Tseng ZL, Ruan JH. Enhanced efficiency of $MAPbI_3$ perovskite solar cells with FAPbX3 perovskite quantum dots. Nanomaterials. 2019 Jan; 9(1): 121.
3. Yuan J, Bi C, Wang S, Guo R, Shen T, Zhang L, Tian J. Spray-Coated Colloidal Perovskite Quantum Dot Films for Highly Efficient Solar Cells. Advanced Functional Materials. 2019 Dec; 29(49): 1906615.
4. Sitapure N, Epps R, Abolhasani M, Kwon JS. Multiscale Modeling and Optimal Operation of Millifluidic Synthesis of Perovskite Quantum Dots: Towards Size-Controlled Continuous Manufacturing. Chemical Engineering Journal. 2020 Dec 7: 127905.
5. Peng Z, Doroodchi E, Evans G. DEM simulation of aggregation of suspended nanoparticles. Powder Technology. 2010 Dec 10; 204(1): 91-102.
6. Wang W, Zhang Y, Wu W, Liu X, Ma X, Qian G, Fan J. Quantitative modeling of self-assembly growth of luminescent colloidal $CH_3NH_3PbBr_3$ The Journal of Physical Chemistry C. 2019 Apr 28;123(20):13110-21.
7. Xu T, Lam ML, Chen TH. Discrete Element Model for Suppression of Coffee-Ring Effect. Scientific Reports. 2017 Feb 20;7(1):1-0.

624233 Findings and Conclusions from a Mobile Worker and Augmented Reality Enabled Continuous Manufacturing Skid Project.

Abstract:

This abstract presents the final conclusions regarding a project initially presented at AIChE 2021 Spring Meeting as an oral presentation (abstract#618949) and poster (abstract#623026).

Background:

Manufacturing USA was created in 2014 to secure U.S. global leadership in advanced manufacturing by connecting people, ideas, and technology. Manufacturing USA institutes convene business competitors, academic institutions, and other stakeholders to test applications of new technology, create new products, reduce cost and risk, and enable the manufacturing workforce with the skills of the future (www.manufacturingusa.com).

MxD encourages factories across the United States to deploy digital manufacturing and design technologies, so those factories can become more efficient and cost-competitive (www.mxdusa.org).

Project detail:

This abstract relates to a continuous process manufacturing skid that has a focus on mobile workers and demonstrates the future of manufacturing by using a digital twin enabled physical skid build. The question the skid answers is "How would you visualize sensor data with the best User Interface/User Experience design to get a seamless experience in continuous manufacturing?"

The findings and conclusions come from a project team consisting of an government owned site personnel, a private chemical processing company, a multi-national technology supplier and multiple service providers. Equipment and assets applied in the project go from sensors to process equipment to pumps/drives to valves to a DCS to a MES and complete engineering and design suite and finally to an IT/OT-integrated and IIoT-enabled setup to support day-to-day operations and mobile workers. Cloud platform with Artificial Intelligence / Machine Learning capabilities is connected to the site and the skid.

Currently there is considerable work done in chemical industry to study the current state and possible future states when it comes to skids, processes, plants, companies, professionals, workers and value chains. Applying novel techniques into operations is in practice an efficient way for human beings to get a real grasp on what is impacted, how it is impacted and what do we want to do with the novel capabilities that come from innovation areas such as the wider digital enterprise or industry 4.0 concept. Concrete studies like this one bring important contributions to the field, especially when introduced as an example in a discussion or presentation at an event such as the AIChE Annual Meeting.

The subject of this abstract was originally planned and divulged in 2019, moved into a contractual obligation in 2020 and is scheduled to conclude in summer of 2021. This timeline allows the authors to go through the whole lifecycle from problem statement to selected methods to real physical results and finally to discuss and expand the implications for future ways of working.

The audience will learn about augmented reality and digital twin use cases around a continuous process testbed. The audience will understand that this skid is available for industry professionals and students to visit at a public facility on Goose Island, Chicago.

Use Cases:

The goal of the process industry test bed skid (physical) and its digital twin (virtual) is to enable field operators, engineers, factory managers, and other key manufacturing roles to gain data-driven insights for increased operational awareness in order to better solve manufacturing problems and optimize processes. A factory manager should be able to communicate with and assign tasks to the mobile worker and view all data necessary to make informed decisions about the assignment of tasks. The solution will be demonstrated using the process test bed and should be agnostic to sensor type, analytics solution, and control type. Visualization of this data should include, but is not limited to, dashboards for factory managers and augmented reality technology for field operation using mobile devices (e.g. tablet and smart glasses).

Conclusions: The three authors of the abstract will each share their findings from their viewpoint, those being the viewpoints of an owner/operator company, the production site and the main technology provider.

632817 Practical Issues in Cybersecurity: From Encryption to Images

Abstract:

Cybersecurity has gained interest for chemical processes in recent years. For example, [1] discusses techniques for considering cybersecurity during traditional chemical engineering procedures such as hazard and risk analysis. One of the main drivers for the cybersecurity discussion is an increased interest in recent years in concepts such as Industry 4.0 [2] and the Industrial Internet of Things [3], and a general increase in data analytics and artificial intelligence approaches. Future work in this domain would benefit from research that enables a more integrated approach to control system cybersecurity in the chemical engineering research field, in which the cyber defenses at a plant are more explicitly considered, as well as advanced control frameworks and concepts. These concepts include resilient frameworks, to analyze how a system functions as a whole and the types of attacks it remains vulnerable to after both, the theoretical requirements for certain resilient control frameworks as well as, practical communication and networking protocols that need to be accounted for. In addition, there are a wide variety of issues becoming more important to consider from a cybersecurity perspective over time. For example, image analysis is important for certain types of decision-making systems, though image data represents another type of information that could be falsified to manipulate such decision-making [4], and computations on the Cloud, where it may be desired to encrypt not only the data but the computations themselves [5, 6], are additional concerns.

Motivated by the above, this talk highlights multiple concepts in the practical aspects of control system cybersecurity as it interfaces with advanced control. For example, using a case study of a benzene hydealkylation process under distributed Lyapunov-based economic model predictive control [7], we explain the role of encryption protocols in enabling safe transfer of information between distributed units, and the extent to which the encryption might enable attacked controllers to be identified on the network. We utilize a process example of a semi-batch polymerization reactor under model predictive control with uncertainty scenarios [8] and offset-free model predictive control [9] to demonstrate the manner in which disturbance-handling is not equivalent to cyberattack resilience, and we also discuss profit targeting as a key metric in distinguishing attacks and faults. We follow up with a discussion of considerations in image analysis, which form part of the increasingly automated world that is more open to attacks, and verification in that domain, and how such concepts relate to cybersecurity and cyberattack resilience. For example, focusing on the zinc flotation process, in which image-based control is considered today [10], we explore the potential use of software such as Blender for creating a preliminary closed-loop simulation methodology for simulating the zinc flotation process, including both system state evolution under control actions, as well as image generation and processing, along with analysis on the impacts of cyberattacks on image data that determine process characteristics. Finally, we will discuss the role of communication techniques, as well as encryption techniques which facilitate communication with the Cloud, in control for nonlinear chemical process systems.

[1] Cormier, A., & Ng, C. (2020). Integrating cybersecurity in hazard and risk analyses. *Journal of Loss Prevention in the Process Industries*, *64*, 104044.

[2] Reis, M. S., & Kenett, R. (2018). Assessing the value of information of data-centric activities in the chemical processing industry 4.0. *AIChE Journal*, *64*(11), 3868-3881.

[3] Sisinni, E., Saifullah, A., Han, S., Jennehag, U., & Gidlund, M. (2018). Industrial internet of things: Challenges, opportunities, and directions. *IEEE Transactions on Industrial Informatics*, *14*(11), 4724-4734.

[4] Bucci, E. M. (2018). Automatic detection of image manipulations in the biomedical literature. *Cell death & disease*, *9*(3), 1-9.

[5] Darup, M. S., Redder, A., & Quevedo, D. E. (2018). Encrypted cloud-based MPC for linear systems with input constraints. *IFAC-PapersOnLine*, *51*(20), 535-542.

[6] Alexandru, A. B., Morari, M., & Pappas, G. J. (2018, December). Cloud-based MPC with encrypted data. In *2018 IEEE Conference on Decision and Control (CDC)* (pp. 5014-5019). IEEE.

[7] Liu, J., Chen, X., Muñoz de la Peña, D., & Christofides, P. D. (2010). Sequential and iterative architectures for distributed model predictive control of nonlinear process systems. *AIChE Journal*, *56*(8), 2137-2149.

[8] Lucia, S., Finkler, T., & Engell, S. (2013). Multi-stage nonlinear model predictive control applied to a semi-batch polymerization reactor under uncertainty. *Journal of process control*, *23*(9), 1306-1319.

[9] Das, B., & Mhaskar, P. (2015). Lyapunov-based offset-free model predictive control of nonlinear process systems. *The Canadian Journal of Chemical Engineering*, *93*(3), 471-478.

[10] Kaartinen, J., Hätönen, J., Hyötyniemi, H., & Miettunen, J. (2006). Machine-vision-based control of zinc flotation—a case study. *Control Engineering Practice*, *14*(12), 1455-1466.

634112 Keynote Talk: Machine Learning and AI Applications in the Chemical Industry

Abstract:

Industrial chemical processes usually come with abundant data and thus provide huge opportunity to leverage machine learning and AI techniques for process improvement. Over the past decade as computational resources have become readily available at much lower costs, together with rapidly growing availability of cloud services, sensor and analytical technology, the number of AI applications adopted for process monitoring, control and optimization have also increased exponentially. Aside from traditional multivariate analysis that is often used for monitoring and fault diagnosis, other analytics fields such as (but not limited to) image analysis, deep learning, sensor fusion, text mining, reinforcement learning, hybrid and physics informed surrogate modeling are also being actively applied to increase reliability, reduce safety events, predict the onset of potential equipment failures and enable real time optimal decision making. This talk will provide examples of AI applications using industrial case studies and also address current challenges as well as outlook.

634113 Keynote Talk: Deploying AI for Automated Monitoring of Physical Infrastructure

Abstract:

This presentation will discuss how AI can be used for automated monitoring of physical infrastructure across various assets and processes. Key topics include:

The purpose and benefits of following an Op-Specific approach to digital transformation based on survey results.

How an operational data platform can reduce resource consumption, improve sustainability goals, and increase revenue.

Steps to implement a specific use-case.

An Operational Data Platform (ODP) can connect the constantly changing realities of the physical world with the power of intelligent software. Plutoshift has built an ODP to unlock the potential of our operational data to enable process monitoring and remote decision making. Plutoshift serves a variety of industries such as chemicals, food & beverage, manufacturing, water treatment, and more. The Plutoshift platform has enabled significant cost reductions ranging from 20-30%. Along with that, it has enabled millions of dollars of revenue increases for customers. The platform is built to help companies achieve sustainability targets as well. For example, the platform helped one global brewer's facility save close to 16 million gallons of water in a single year. These savings are being driven by enhanced visibility of assets and processes across the physical infrastructure.

630200 Resilience in the Face of COVID-19: Managing a User Program at the Molecular Foundry

Abstract:

Introduction

The Molecular Foundry is a Nanoscale Science Research Center, located at Lawrence Berkeley National Laboratory (Berkeley Lab) in Berkeley, California, sponsored by the Department of Energy (DOE). The Molecular Foundry is a knowledge-based User Facility, where leading-edge resources in instrumentation, analytical capabilities, and scientific expertise in nanoscale science are made available via a peer-reviewed proposal process. Roughly 1000 researchers or "users" come to the Foundry from around the world to conduct their work each year. The COVID-19 pandemic presented challenges and opportunities in conducting collaborative on site research. This presentation focuses on the operation of the user program amid the pandemic, with an emphasis on strategies for maintaining research productivity and user access. We also present an outlook on ways that gains in virtual collaboration and science during the pandemic can provide opportunities in a post-pandemic world.

The Molecular Foundry provides scientific expertise and capabilities distributed among seven facilities: Inorganic, Organic, and Biological facilities for synthesis, preparation, and assembly; Nanofabrication, for processing and integration; the National Center for Electron Microscopy and Imaging and Manipulation, for characterization; and Theory, for understanding and predicting material properties.

The Molecular Foundry maintains a highly collaborative research model. Staff scientists and technical staff at the Molecular Foundry interact with users through instrument training, consultation during experiments, support with data interpretation, and collaborative writing of journal articles. For each 12 month project, users stay in the local area for an average of three months while they work on site.

Spring 2020 Shelter-in-Place

In spring 2020, the scale and urgency of the COVID-19 pandemic in the United States became apparent, and the city of Berkeley enacted a Shelter-in-Place order effective March 17, 2020. The Molecular Foundry responded by pausing research operations on site while keeping research facilities safe and ready for reopening. This directly impacted every one of the hundreds of research projects that were active at the time, and presented a challenge to the Molecular Foundry's commitment to safety and user access.

The Foundry staff collaborated to respond to this challenge with a holistic approach to keep science moving forward safely. All user projects received a 12-week extension, to account for this pause on work. During this period, the Foundry developed protocols for safe work on site, aligned with requirements for Berkeley Lab and applicable regional, state, and national guidelines. These included the use of face coverings, hand-washing, distancing/density control, and surface disinfection.

While maintaining a very low density of people on site, a few high-priority projects were allowed to proceed, including research on COVID-19 and SARS-CoV-2 detection, treatment, and transmission mitigation. These early research activities served as a testing ground and model for the further expansion of access in the months to follow. Staff from across all seven facilities began to explore investment in virtual modes for training, user support, and instrument access amid the pandemic. Berkeley Lab developed new training modules for safe work on site during the COVID-19 pandemic, launched a lab density study to determine appropriate safe occupancy limits for all laboratories across the site, and developed a Health Pledge,symptom check system, and a specific COVID19 training module that was required for all on site personnel.

Summer 2020 and Beyond

The months that followed the onset of the pandemic were characterized by continual adaptation to the emerging pandemic. As of the writing of this abstract, this phase continues to evolve as vaccination campaigns are gaining momentum nationally. These are some of the strategies employed to manage the user program during this phase of the pandemic:

Communication

Communication with the user community and the staff of the Molecular Foundry was a central pillar of maintaining safe work at the Molecular Foundry. This included email messages to all users generally, frequent discussions among staff across all roles, and communication with Berkeley Laby leadership.

Shift Work to Manage Density

Upon emerging from the full shelter-in-place order, moderate numbers of researchers were allowed to work on site at the Molecular Foundry. The total number of persons accessing Berkeley Lab was managed centrally by the Laboratory, and the Molecular Foundry worked with an allotment of "slots" for site access throughout the duration of the pandemic restrictions. A three-shift model was deployed. Slots were distributed among the seven facilities, and researchers requested a slot ahead of time, subject to approval by Molecular Foundry leadership and compliance with the Health Pledge, weekly symptom check, and a specific COVID19 training requirements.

Prioritizing Research Projects

Facing limited access compared to normal operations, it was necessary to consider carefully which staff and user work could be accomplished on site. The Molecular Foundry values equity of access for users, held in balance with safety precautions. We aimed to carefully consider such factors as the potential to impact the fight against COVID-19, DOE and Berkeley Lab research priorities, inherently safe activities, experimental complexity, travel considerations, and user training.

Badged Entry for Approved Work

During the pandemic, the Molecular Foundry implemented badge-reader entry for all buildings. Coupled with weekly calendars and daily posting of all persons with prior approval to enter the buildings, this provided a layer of accountability around access to the spaces.

Regular Safety Meetings

Berkeley Lab implemented weekly virtual meetings focused on COVID-19 safe operations and other safety and space planning topics. These meetings require attendance by at least one staff representative from each of the seven facilities, and are open to all Molecular Foundry staff and users. Each facility hosted a daily safety check-in for work planning.

Carefully Considered Training

Training presented a particular challenge during the pandemic, as many users receive training on complex tasks at the Molecular Foundry, traditionally accomplished side-by-side. After several months, the Foundry formalized guidelines to consider for new user training, considering such factors as lab occupancy, the potential to develop tools to support distanced training, inherent hazards in the proposed work, and robustness of the instruments.

Development of New Remote and Virtual Training and Access

Molecular Foundry staff rallied around the call to support user access virtually. Specific responses are adapted to the needs and opportunities of each facility. A sampling of tools include: training videos, holographic augmented reality for training, audio headsets to support distancing, remote operation of select equipment, remote login for data processing, virtual presence via video conferencing during experiments, telepresence robots, and a variety of informal communication and collaboration tools.

Calls for Proposals Scope

Two of the standard calls for proposals have been fully within the pandemic. While we kept the calls open for the most exciting research across nanoscience, we balanced the practical considerations with some emphasis for these calls. We encouraged users to submit proposals that included COVID-19 research, theoretical and computational work, remote access to instruments, or data collection for remote analysis, and proposals requesting samples of materials developed at the Foundry.

Expanded Definition of Remote Users

The Molecular Foundry is one of five Nanoscience Research Centers sponsored by the DOE, and therefore shares certain reporting parameters for our user programs. Effective for FY 2020 reporting, the centers collectively adopted an expanded definition of a remote user, to recognize the increasing role of remote access for science at these knowledge-based user facilities.

Virtual Meetings and Conferences

Meetings have shifted online, supported by zoom video conferencing. This includes smaller team meetings, larger seminars spanning the Molecular Foundry, and our annual User Meeting--a nanoscale science research conference which drew 900 registrants in 2020.

Outcomes and Future Outlook

The pandemic presented challenges to maintaining a strong nanoscience research user program. Nevertheless, overall metrics reflect resilience in the face of these challenges. The User Program remains strong, serving 643 on-site and 97 remote users in FY 2020, according to formal reporting definitions, with a total of over 1500 named researchers on active projects during FY 2020. Demand for the user program remains strong, with around 225 standard proposals received during the most recent spring 2021 call for proposals, roughly two thirds our pre-COVID maximum. Research productivity as measured by journal articles has remained robust when comparing 2020 to 2019 and 2018.

The shift to increased virtual access presents opportunities to expand the geography of our user base to include more researchers outside our local area, due to reduced barriers in travel once post-pandemic travel resumes. We have seen an increase in sample-only proposals, enabling new ways of working together across institutions. We also have potential to include more people in the active experimental data collection phase, as undergraduate researchers and principal investigators increasingly log in remotely to participate in experiments that previously were completed by just one member of the research team. Furthermore, shifting our annual User Meeting online has facilitated a tripling of conference registrations, with potential to reach new communities with our science.

As we look ahead to 2022 with optimism, we hope to learn from the unique challenges presented by the pandemic. We will have new facets of this experience to manage, including a safe return to work for our users. The Molecular Foundry remains committed to advancing user science to bring solutions to the world.

629703 Global Supply Chain Optimization for COVID-19 Vaccines Under Covax Facility

Abstract:

With the pandemic in action, over 130 million people across 219 countries have been infected by COVID-19. The number of mortalities due to COVID-19 has climbed to around 3 million [1]. Though around 80% of the infected patients have recovered, several patients are becoming reinfected recently appeared. As a result, efforts for COVID-19 vaccines have accelerated around the world. The world's top pharmaceutical companies and research organizations have developed multiple vaccines. It is expected that the transmission can be slowed considerably if a sufficient population is vaccinated. Though a typical vaccine takes around ten years for approval, several COVID-19 vaccines have already been developed and approved in record times. As of 31 March 2021, around 750 million people worldwide have been vaccinated [2]. With world governments aggressively supporting vaccine developments and procurements, the vaccination drive is progressing rapidly.

However, to end the pandemic, we not only need COVID-19 vaccines, but the development of vaccines must be followed by a fair, fast, and efficient global distribution. The current challenge is the timely delivery of the vaccines. The expected volume of the vaccines per month adds up to around 850 tons and is more than the UNICEF suppliers' capacity combined [3]. Two main concerns regarding the global COVID-19 vaccine distribution are 1) Limited supply of resources for the delivery of the vaccines, e.g., cold chain equipment, and 2) Vaccine nationalism, i.e., governments signing agreements with companies to procure surplus vaccines. To overcome these challenges, the COVAX facility was launched by WHO for multiple vaccines to prevent vaccine nationalism. It aims to ensure fair access to COVID-19 vaccines for all countries, irrespective of their financial status. It allows countries to pool resources for effective vaccine development and distribution. Under COVAX, deals for vaccine procurement have already been made. However, vaccine distribution is still under development. It faces various challenges due to (1) the variety of vaccines, (2) person-specific doses, 2) unknown country-specific demands required for herd immunity threshold, especially low-middle income countries, (5) low-temperature storage protocols for most vaccines, and (6) unanticipated disruptions in both production and distribution.

Addressing these challenges requires a multi-prong approach that must go beyond straightforward pharmacological measures. The two primary goals are to (1) estimate the herd immunity threshold and (2) effect an efficient global vaccine distribution. In this work, we explore the time-varying nature of reproduction number during the first wave of the

COVID-19 pandemic using confirmed cases, recovered cases, fatalities, and timings of lockdown phases. We propose a mean-field epidemiological model for COVID-19 that extends the classical SEIR model [4]. We describe NEIASCFR, an 8-compartmental model with variable transmissibility over time, by considering the time-varying nature of parameter values over different lockdown phases. We estimate our model's parameters with public data on confirmed cases, recovered cases, and fatalities for each country. The parameter values are used to determine reproduction number and herd immunity threshold. The model is applied to study the herd immunity threshold of 9 countries (India, Nigeria, Iran, Pakistan, Thailand, Mexico, Democratic Republic of Congo, Egypt, Thailand), constituting 50% of demand under the COVAX facility.

Using this within the framework of COVAX, we develop a global supply chain model to optimize the production and allocation of vaccines to participating countries from various production centers. The supply chain was optimized considering potential disruptions, vaccine approval dynamics, and vaccine candidates' approval with different risk mitigation strategies. Given: 1) Network configuration consisting of production facilities and different regions. 2) Initial vaccine procurement quantities. 3) Capacity of different storage containers. 4) Costs of procurement, storage, transportation, and risk mitigation per unit. 5) Different scenarios, we determine 1) Production quantities of vaccines at different centers and vaccine allocation to each demand region. 2) Extra procurement of vaccines at each center. 3) Cost of procurement and transportation.

References:

1. Worldometer. COVID-19 CORONAVIRUS PANDEMIC. 2021; Available from: https://www.worldometers.info/coronavirus/.

2. Bloomberg. Covid-19 Tracker. 2021; Available from: https://www.bloomberg.com/graphics/covid-vaccine-tracker-global-distribution/.

3. UNICEF, UNICEF outlining plans to transport up to 850 tonnes of COVID-19 vaccines per month on behalf of COVAX. 2020.

4. Li, M.Y. and J.S. Muldowney, Global stability for the SEIR model in epidemiology. Mathematical biosciences, 1995. 125(2): p. 155-164.

632079 Sustainability and Resiliency Assessment for Industrial Synergies between Renewable Energy and Chemical Production

Abstract:

Photovoltaic (PV) is the fastest growing energy technology and is expected to maintain this growth in the coming years. Due to the durability of PV modules, there is a lengthy time gap between installation and decommissioning of the modules, and it is expected to create large scale waste worldwide in the near future. On the other hand, chemical manufacturing systems are undergoing significant transition towards bio-based production to meet the goal of sustainable low carbon feedstock. This work presents a potential synergy between renewable energy wastes and bio-based polymer production through extending a previously proposed system to circularize the PV lifecycle by diverting organic byproducts from PV recycling to agricultural production. We identify and introduce additional synergies as links in the network to close the loop between PV module disposal and manufacturing. The chemical outputs from the PV recycling process are used as fertilizers for crops that are potential feedstock for biopolymer production. These biopolymers are used in the manufacturing of PV backsheets, replacing the current petroleum-based products. We quantify the total life cycle environmental and economic benefits at scale of implementing this synergy along with total reduction in GHG emissions. To evaluate the design, we also use a network-based approach to understand the structure and topology of the proposed system by exploring various network measures as well as information theory-based measures, to assess resilience and long-term sustainability. We show quantitatively that network measures like diameter and density are ineffective at scoring the network resilience due to the nature of these measures in original network science literature, while centrality measures are useful for finding the relative importance of certain nodes, which may highlight network vulnerabilities that factor into the overall resilience. Likewise, ascendency and reserve values for the network appear useful in determining the overall resilience and sustainability of the network, however, based on their application to comparing two alternative network scenarios, these metric interpretations should be more rigorously defined for industrial networks before being incorporated for industrial network design.

633441 A Community-Level Energy Planning Model for the Dynamic Optimization of Waste-to-Gas and Waste-to-Power Anaerobic Digestion Supply Chains.

Abstract:

Society has growing concerns over the long-term environmental impacts of climate change and the accumulation of organic waste in local communities. Community energy planning is increasingly focused on the use of novel technologies to provide sustainable and cost-effective heat and power. Furthermore, changes in energy demand patterns support the development of dynamic optimization models of energy dispatch. This study focuses on the use of anaerobic digestion (AD) technology for converting human, agricultural, and industrial waste into renewable natural gas (RNG) and power. We develop a dynamic mixed integer linear programming (MILP) in Calliope to optimize the production and dispatch of RNG and power.
Anaerobic Digestion (AD) is a microbiological process that is used in processing organic, biodegradable wastes and side streams into methane-rich biogas and nutrient-rich digestate. Previous techno-economic (TEA) and lifecycle (LCA) analyses indicate that AD can be a commercially competitive and sustainable technology for communities to generate clean energy. In this study, we evaluate the use of AD in Ames, Iowa to convert human and animal waste, biomass, and wastewater into RNG and power. RNG and power are consumed locally to provide industrial, commercial, and residential heat and power demand. The AD process generates by-product digestate that is returned to Iowa farms. The model covers simulates 1 year of energy demand in hourly and monthly increments.

The MILP identified 13 locations in the vicinity of Ames, Iowa capable of producing power or RNG. The electricity price ranged between \$0.065 - \$0.18/kWh and the RNG price is estimated at \$0.026/kWh (\$7.62/mmbtu). Electricity production varied between 30 and 50 million kWh per month. Gas production varied between 15 and 40 million kWh per month. All the electricity demand in Ames, Iowa could be supplied from local waste resources. Summer gas demand can be supplied locally. However, the gas supply will need to be supplemented with gas from the external grid to meet winter demand. Technological improvements and public incentives could increase the adoption of AD in community energy planning systems. Further research will evaluate the impacts of variabilities in energy supply and demand to address grid disruptions.

625575 Asset Integrity Management (AIM) for Hydrogen: Major Incident Risk Reduction By Integrating a Risk-Based Safety Approach

Abstract:

Asset Integrity Management (AIM) is a well-understood process for optimizing the management of assets over their whole life-cycle. If applied correctly, AIM can offer asset owners and operators the ability to manage risk and enable the assertion, with confidence on an evidential basis, that their assets are safe and reliable. For AIM to be most effective, the assets' design must consider the whole life-cycle of the assets' operation. Adopting a risk based approach to safety and operations as early as possible in the design phase of a facility is demonstrated to improve long-term reliability, availability and maintainability. Combining a risk based approach at the design stage with AIM during operations has the potential to simplify and align processes. Additional benefits include the improvement of long-term reliability, availability and maintainability. By integrating a risk based approach with AIM the realization of common goals can be facilitated: the reduction of major accident hazard risks and consistent performance of assets. For a sector on which many emergent national decarbonization policies are reliant upon to deliver "net zero", an all-encompassing risk based safety and AIM approach offers asset owners and operators in the beginnings of the hydrogen economy boom a means by which to deliver business objectives profitably without major incident.

625565 More Haste, Less Speed: Proportionate Supply Chain Risk Management for the Accelerating Hydrogen Economy

Abstract:

The long awaited acceleration of the hydrogen economy is dawning. It is viewed as a necessity in the race to decarbonize industry, heat and transport, amongst other key sectors. Existing hydrogen economy supply chains serving the current markets face the challenge of ramping up to meet future demand. Aside from hydrogen growth, many organizations are increasingly concerned about supply chain risk, not only the traditional supplier issues of late delivery and poor quality but also other risks with security, environmental, social and ethical impacts. Furthermore, hydrogen as an energy vector requires significant safety management to mitigate its potential to cause major accidents. The confluence of increasing demand, major accident hazards associated with hydrogen and concerns of supply chain risks present a significant challenge for those organizations in the hydrogen supply chain. When viewed through the lens of rapid demand acceleration, supply chain risks are magnified by the potential for short-cuts and risk taking. Such risks are exacerbated by the potential set-back in the global decarbonization goal, should a major accident be realized. Formal Supply Chain Risk Management (SCRM) techniques adopting a risk based approach are explored to help mitigate supply chain risks in a proportionate way, reducing supply chain vulnerability, ensuring continuity if risks occur and increasing value within the supply chain through improved performance. Crucial attributes of successful SCRM are considered to be: application of good governance, early supplier engagement, proportionate assurance levels based on risk, supply chain mapping, monitoring of risk events/risk event pre-cursors and risk register maintenance.

637411 The Hard Skills of Emotional Intelligence for Effective Engineering Leaders

Abstract:

High performing teams and leaders with emotional intelligence are more likely to achieve their goals. This workshop featuring Dr David Caruso of Yale's Center for Emotional Intelligence, focuses on the ability model of emotional intelligence and begins with a hands-on illustration of the four "hard" skills of EI, followed by demonstrations of the notion that "emotions are data" and the cognitive costs of emotion suppression. After a review of a brief case study, we then provide participants with a means to apply these skills to situations faced by leaders in engineering using the Emotional Intelligence Blueprint.

AUTHOR INDEX